宝宝嘻哈乐学丛书

 数学新玩法

罗国庆 陈良萍 编著

U0238422

我的宝贝
——To My baby Dear ____ My baby ____

山东大学出版社

图书在版编目（CIP）数据

数学新玩法 / 罗国庆，陈良萍编著 .
—济南：山东大学出版社，2014.10
（宝宝嘻哈乐学丛书）
ISBN 978-7-5607-5069-9

Ⅰ . ①数…
Ⅱ . ①罗… ②陈…
Ⅲ . ①数学—儿童读物
Ⅳ . ① O1-49

中国版本图书馆 CIP 数据核字（2014）第 149802 号

策划编辑：刘森文
责任编辑：刘森文 郑琳琳
封面设计：祝阿工作室

出版发行：山东大学出版社
  社 址：山东省济南市山大南路 20 号
  邮 编：250100
  电 话：市场部（0531）88364466
经 销：山东省新华书店经销
印 刷：济南新先锋彩印有限公司
规 格：880 毫米 ×1230 毫米 1/16
    4 印张   53 千字
版 次：2014 年 10 月第 1 版
印 次：2014 年 10 月第 1 次印刷
定 价：18.00 元

认识数量 — 选择正确的数字

请圈选出下面每一组中与物体数量匹配的数字。（Draw a circle around the correct number of objects in each group.）

3　4　5

6　7　8

2　3　4

4　5　6

6　7　8

8　9　10

8　9　10

5　6　7

1

数学迷宫 — 偶数路径

这只小兔子只有通过偶数的方块才能吃到胡萝卜。请帮助这只小兔子从下面格子最底行蹦到最顶行吧，只能蹦到有偶数的方块中。你可以往上、下、左、右或往斜对角蹦，要注意每次只能蹦一个方块。

（Help the bunny hop from the bottom row of the grid to the top row by jumping only on squares that have an even number. You can move up, down, sideways, or diagonally, but only one square at a time.）

温馨提示：以 0、2、4、6、8 结尾的数字是偶数。

终点

20	5	31	99	25	9
47	12	100	13	11	49
1	33	5	22	88	21
19	21	23	3	35	50
5	49	32	60	4	67
33	6	25	13	55	9
13	70	15	7	47	14
35	3	40	14	17	3
10	9	5	15	8	11
13	7	31	41	2	23

起点

2

推理物品 — 数字

根据线索，对每个不符合的数字打叉，将剩下的一个正确的数字圈选出来。（Follow the clues and cross out each number that doesn't fit until you are left with the correct number. Circle it.）

线索 1：这个数字比 3 大。
（Clue 1: The number is greater than 3.）
线索 2：这个数字比 8 小。
（Clue 2: The number is less than 8.）
线索 3：这个数字不等于 2 加 2。
（Clue 3: The number does not equal 2 + 2.）
线索 4：这个数字不等于 8 减 1。
（Clue 4: The number does not equal 8 - 1.）
线索 5：这个数字不等于 4 加 2。
（Clue 5: The number does not equal 4 + 2.）

9 6 2 8 5 1 3 7 4 10

色彩缤纷 — 给数字和字母涂色

请根据下面的线索来给每个数字和字母涂色。（Color each number and letter according to the following clues.）

0= 灰色（gray）；1= 深绿色（dark green）；2= 浅绿色（light green）；3= 黄色（yellow）；4= 浅蓝色（light blue）；5= 橙色（orange）；6= 紫色（purple）；7= 红色（red）；8= 粉红色（pink）；9= 棕色（brown）；A= 深蓝色（dark blue）；B= 红色（red）；D= 橙色（orange）；G= 紫色（purple）；H= 粉红色（pink）；K= 黄色（yellow）；P= 棕色（brown）；T= 浅蓝色（light blue）；Z= 黑色（black）

4

认识数量 — 填写数字

通过填写下面缺少的数字来练习数数。（Practice counting to one hundred by filling in the missing numbers below.）

1	2					7			
	12				16				
		23						28	
					36				
			44						50
								59	
				65					
71							78		
		83					87		
			94					99	

认识数量 — 机器人的数量

回答下面问题：

有多少机器人是由椭圆组成的？（How many robots are made of ovals?）
有多少机器人不是由椭圆组成的？（How many robots are not made of ovals?）
有多少机器人是由矩形组成的？（How many robots are made of rectangles?）
有多少机器人不是由矩形组成的？（How many robots are not made of rectangles?）
有多少机器人是由三角形组成的？（How many robots are made of triangles?）
有多少机器人不是由三角形组成的？（How many robots are not made of triangles?）
有多少机器人是由星形组成的？（How many robots are made of stars?）
有多少机器人不是由星形组成的？（How many robots are not made of stars?）
一共有多少机器人？（How many robots are there?）

认识钞票 — 纸币和硬币的面值

请将钱数与匹配的纸币或硬币连线。（Connect the number with the matching bill or coin.）

1元

2元

5元

10元

20元

50元

100元

5角

2角

1角

5分

1分

硬币涂色 — 发现财宝

将寻找财宝的路径上的硬币按下面要求涂色。（Color the coins in the path to the treasure.）
一元硬币 (1 yuan coins)：红色 (red)；　　五角硬币 (0.5 yuan coins)：金黄色 (golden yellow)；
一角硬币 (0.1 yuan coins)：绿色 (green)；　五分硬币 (0.05 yuan coins)：蓝色 (blue)。

图中寻宝 — 找元宝

你能找出多少个元宝呢？（How many gold ingots can you find in the picture below?）

9

趣味数数 —— 以 2 递增或递减的方式数数

请按照以 2 递增的方式从 10 数到 50，将数字写在泡泡里。（Count forwards by twos from 10 to 50, and write the numbers in the bubbles.）

请按照以 2 递减的方式从 80 数到 58，将数字写在花朵中。（Count backwards by twos from 80 to 58, and write the numbers in the flowers.）

与众不同 — 数字积木

下面有一幅图与其他图不同，请将这幅图找出来。（Find the picture which is different from the others.）

二分之一——分成两半

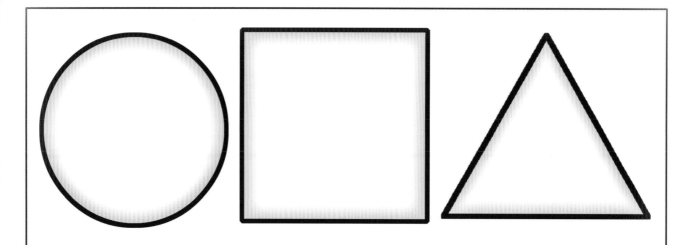

将每个图形分成两半。 （Divide each shape in half.）

将圆形的一半涂成红色。 （Color half the circle red.）

将正方形的一半涂成蓝色。 （Color half the square blue.）

将三角形的一半涂成绿色。 （Color half the triangle green.）

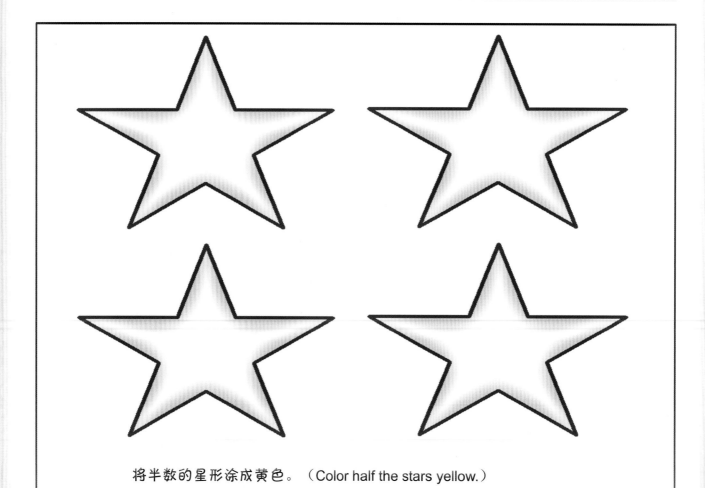

将半数的星形涂成黄色。 （Color half the stars yellow.）

看懂时钟 — 模拟时钟与数字手表

请将显示时间匹配的模拟时钟与数字手表连接起来。（Connect the analog clocks to the digital watches according to the same time indicating.）

拼接图片 — 涂写数字

请将下面图片的序号填在相对应的方框里，拼接成一幅完整的图像。（Fill the numbers in the blanks to make a complete picture.）

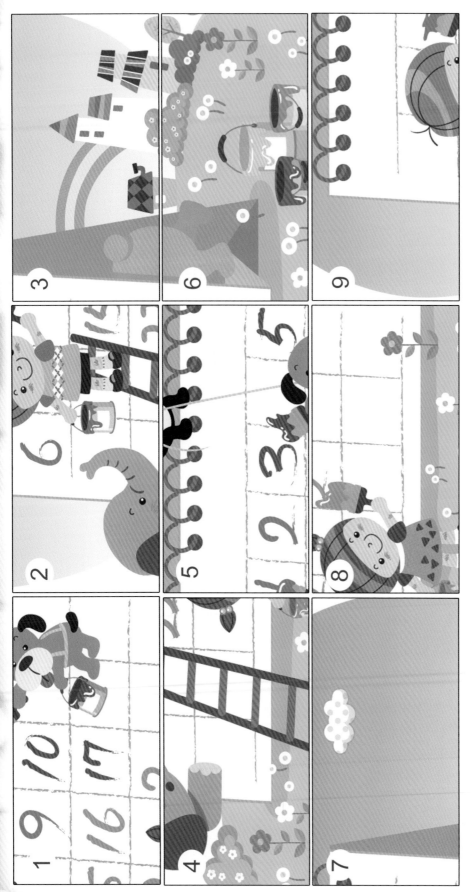

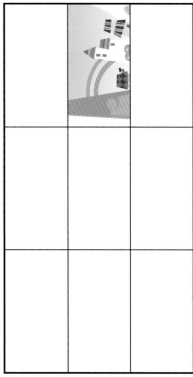

趣味数数 — 以 5 递增或递减的方式数数

请按照以 5 递增的方式从 5 数到 80，将数字写在泡泡里。（Count forwards by fives from 5 to 80, and write the numbers in the bubbles.）

请按照以 5 递减的方式从 100 数到 45，将数字写在花朵中。（Count backwards by fives from 100 to 45, and write the numbers in the flowers.）

连接数字 — 夏天捕蜻蜓

按照从 1 到 50 的顺序连线，同时说出数字。（Draw a line from 1 to 50 in order while saying each number.）

你能发现蜻蜓吗？（Have you found the dragonfly?）

培养财商 — 钞票面值的加减

请在空白处填上正确的数字。（Fill in the blanks with the proper numbers.）
温馨提示：请让爸爸妈妈帮你一起完成。

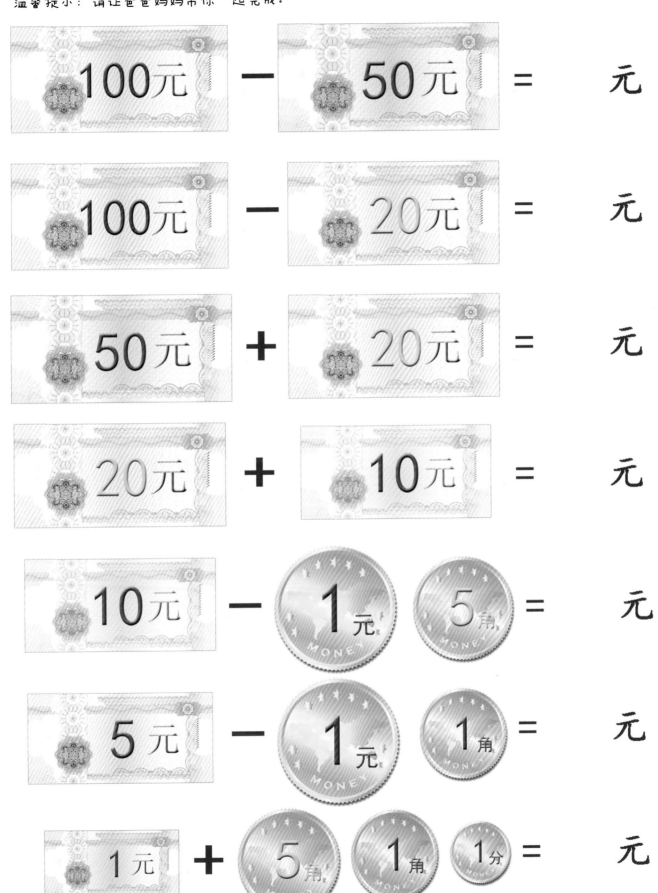

100元 — 50元 = ＿＿元

100元 — 20元 = ＿＿元

50元 + 20元 = ＿＿元

20元 + 10元 = ＿＿元

10元 — 1元 5角 = ＿＿元

5元 — 1元 1角 = ＿＿元

1元 + 5角 1角 1分 = ＿＿元

培养财商 — 储钱罐中的硬币

数一数下面小猪储蓄罐里的钱，并将数量填写在空白处。（Count the coins in the piggy banks below. Write the amounts in the blanks.）

温馨提示：请让爸爸妈妈帮你一起完成.

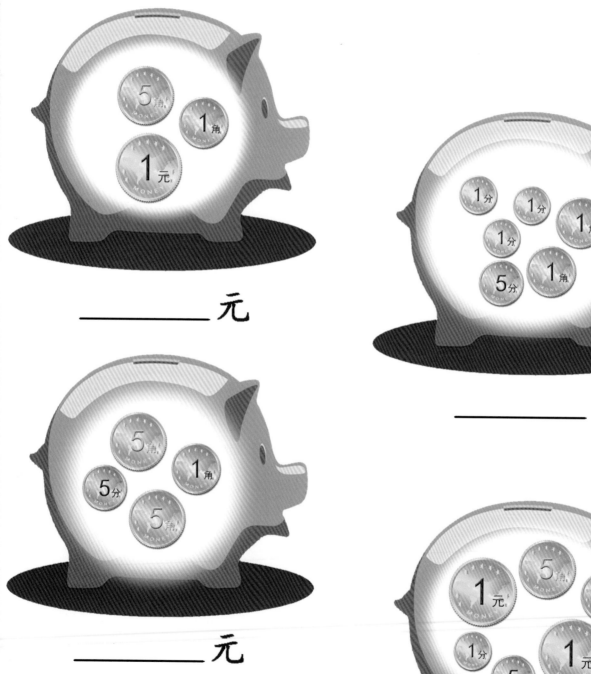

_____ 元

_____ 元

_____ 元

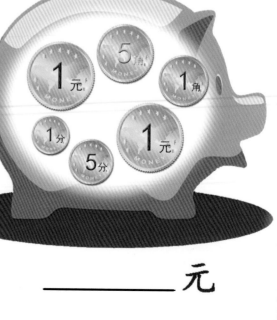

_____ 元

二分之一 — 分给朋友

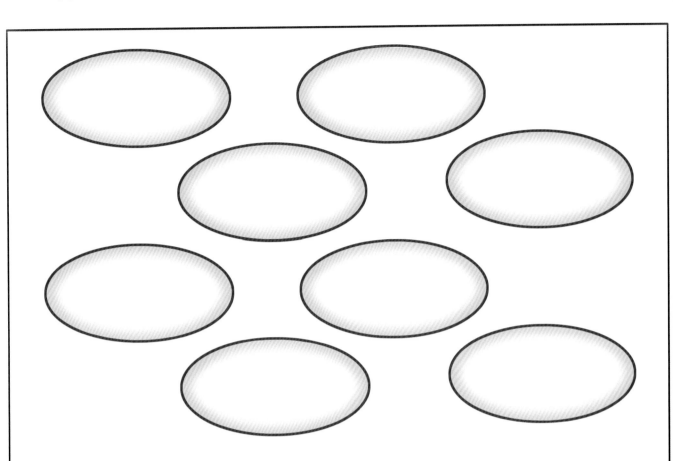

将半数的椭圆形涂成橘红色。（Color half the ovals orange.）

如果你有两块糖，你将一半的糖给了朋友，那么你还剩几块糖？
（If you have two candies and give away half of them to a friend, how many do you have left?）

如果你有四块饼干，你将一半给了朋友，你还剩几块饼干？
（If you have four cookies and give away half of them to a friend, how many do you have left?）

破译密码 —— 电话上的秘密

小雅正在给她的朋友们打电话，邀请他们去游泳。（Xiao Ya is calling her friends to invite them to swim.）

使用下面电话键盘作为解码器，破译出她的谈话内容。（Using the phone keypad below as a decoder, figure out what she is saying.）

温馨提示：请让爸爸妈妈帮你一起完成。

第一个数字表示按键号码，斜线后的数字表示是第几个字母。（The first number tells the number of button. The number after the slash tells the letter position.）

例如，"3/2"表示按键3上字母DEF的第2个字母，即字母"E"。如果数字后没有斜线，表示是按键上的第一个字母。（For example, "3/2" means the second letter on the button 3, that is "E". A number with no slash after it means the letter is in the first place.）

4/3 9-2-6/2-8

8-6/3 7/4-9-4/3-6

2/3-6/3-6-3/2

9-4/3-8-4/2 6-3/2

7/4-3/2-3/2 9/3-6/3-8/2

7/4-6/3-6/3-6/2

连接数字 — 草地上的蚂蚱

按照从 1 到 50 的顺序连线，同时说出数字。（Draw a line from 1 to 50 in order while saying each number.）

草地上是什么？（What is on the grass?）

色彩缤纷 —— 按数字给花朵涂色

根据数字来涂色，就可以看到许多有颜色的花儿了。（Color by number to see the colorful flowers.）
1= 深绿色（dark green）；2= 浅绿色（light green）；3= 黄色（yellow）；4= 橙色（orange）；
5= 紫色（purple）；6= 红色（red）；7= 粉红色（pink）；8= 棕色（brown）；9= 浅蓝色（light blue）

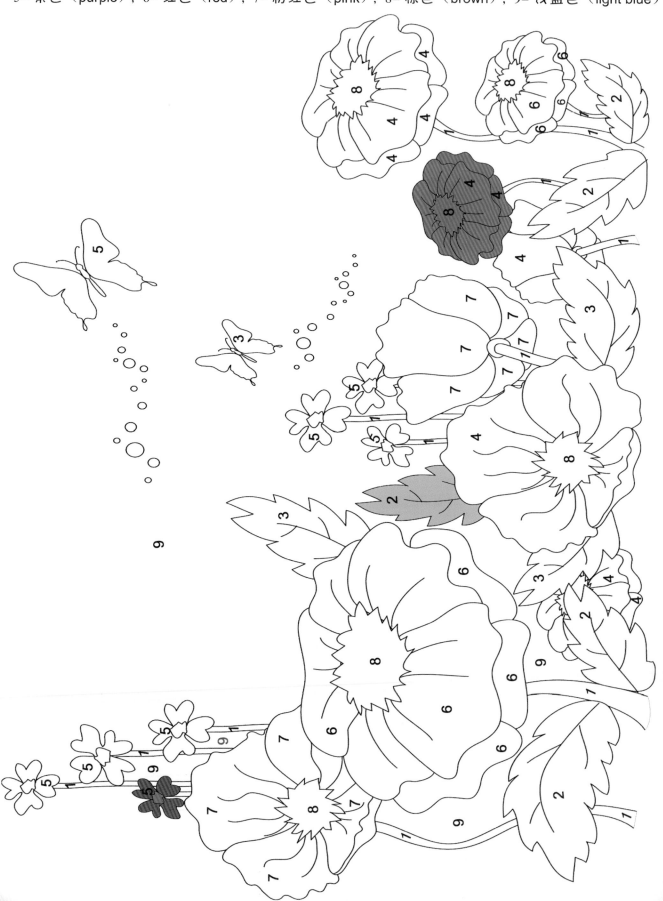

数学游戏 — 六边形上的数字

请用数字 1 到 9 来填充空白圆圈，使每个大六边形的 6 个数字总和为 10 或 20。（Use the numbers from 1 to 9 to fill in the empty spaces. When the puzzle is finished, the six numbers around every big hexagon should add up to 10 or 20.）

温馨提示：请让爸爸妈妈帮你一起完成。

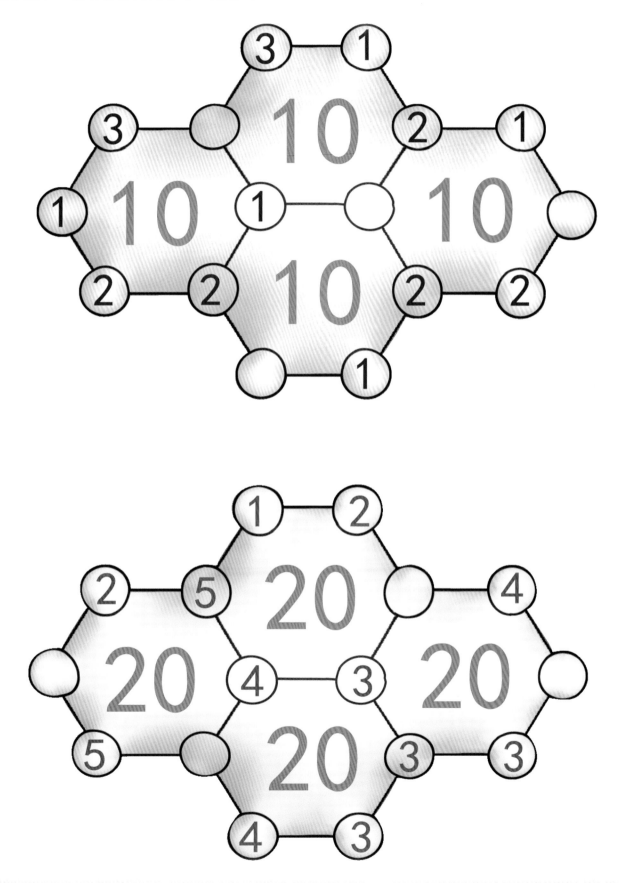

看懂时钟 — 写下时间

请在下面横线上写下钟表所显示的时间。（Please write down the indicated time of each clock on the line.）

温馨提示：请让爸爸妈妈帮你一起完成。

_____ _____ _____ _____

_____ _____ _____ _____

_____ _____ _____ _____

_____ _____ _____ _____

学不等式 — 气球的多与少

符号"＞"表示"大于"；符号"＜"表示"小于"。例如：3>2, 2<4。在下图横线中填入树叶的数量，然后圈选出正确的不等号。（Here, ">" means "greater than", "<" means "less than". For example: 3>2 and 2<4. Fill in the number of leaves, and then circle the correct inequality sign.）

温馨提示：请让爸爸妈妈帮你一起完成。

连接数字 — 花朵

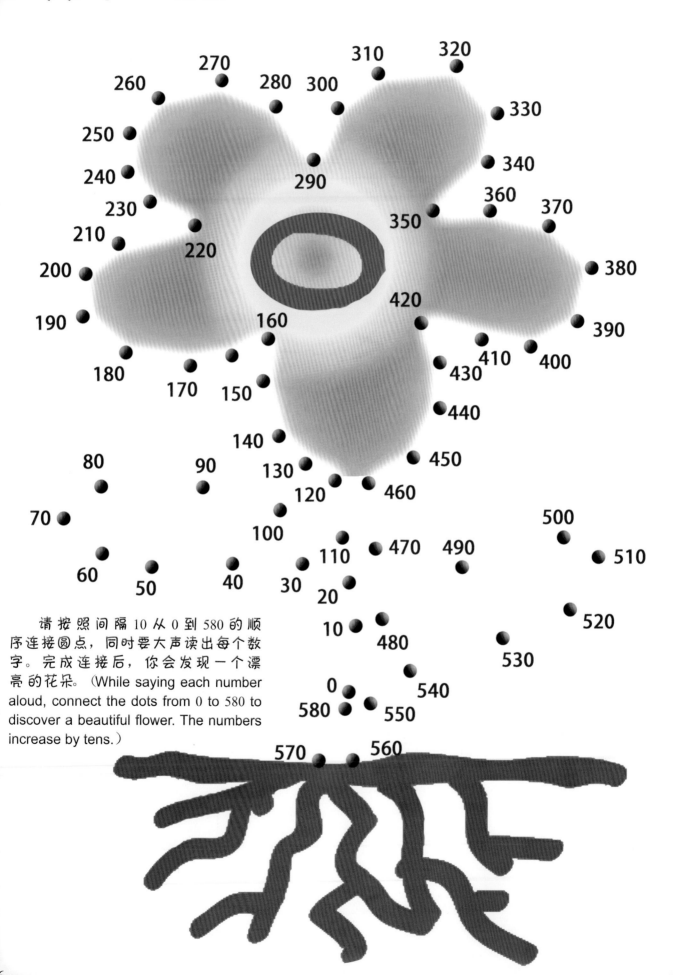

请按照间隔 10 从 0 到 580 的顺序连接圆点，同时要大声读出每个数字。完成连接后，你会发现一个漂亮的花朵。（While saying each number aloud, connect the dots from 0 to 580 to discover a beautiful flower. The numbers increase by tens.）

看图算术 — 学做减法

将答案填写在红框中。（Fill in the answer in the red box.）

		=	
2	- 2	=	
4	- 2	=	
5	- 4	=	
7	- 4	=	

破译密码 — 学做加法

阅读下面的问题，解答加法题。然后用数字与字母对应的关系表，将每个数字用字母代替，最后找出问题的答案。（Read the question, solve the addition problems, and then use the number and letter table to substitute a letter for each number and find the answer to the question.）

温馨提示：请让爸爸妈妈帮你一起完成。

在中国，什么熊是黑白色的？
（What bear is black and white in China?）

数字与字母对应关系

0=e 1=d 2=h
4=p 5=n 6=a
    7=t

3 +4	0 +2	0 +0

答案

字母
解密

3 +1	3 +3	2 +3	1 +0	2 +4

亲子手工 — 制作时钟

需要的材料：图画纸、胶水、剪刀、一个曲别针和铅笔。（Prepare the following materials: construction paper, glue, scissors, a paper clip, and a pencil.）

温馨提示：需要家长协助孩子一起完成。

制作步骤：

1. 剪下下页的时钟的面和指针。（Cut out the clock face and hands on next page.）

2. 将时钟面粘贴在一张有颜色的图画纸上。（Glue the clock face onto a piece of colored construction paper.）

3. 在时钟的表面和指针的中心各打一个小孔，用一个曲别针把钟面和指针固定好。（Make a small hole in the center of the clock's face and one in each of the clock's hands. Use a paper clip to attach the clock's hands to the clock. ）

4. 用这个时钟来练习看时间。（Use this clock to practice telling the time.）

5. 在图画纸上写下一天中你做某件事情的时间，将时钟指针调整到与时间匹配。例如起床时间、上学时间、午饭时间、休息时间、放学时间和睡觉时间。（Write down on the construction paper the times you do certain things during the day and adjust the clock's hands to match each of these times. For example, the time you get up, the time you go to school, lunch time, recess time, the time school is over, and bedtime.）

拼接图片 — 学习加法

在 A、B、C 三张图片中，哪一张图片能完全与下面的拼图吻合呢？（Which piece among A, B, and C fits exactly into the jigsaw puzzle?）

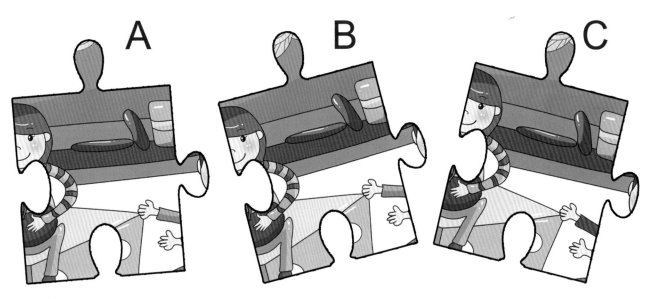

简单加减 — 小狗的食物

请先算出骨头上算式的答案，再与相应的小狗身上的数字连线。（Calculate the answers on the bones and then connect the bones to the correct dogs.）

9-1=___

6+4=___

10-0=___

10-2=___

4+3=___

4+5=___

10-1=___

3+4=___

8+2=___

9-0=___

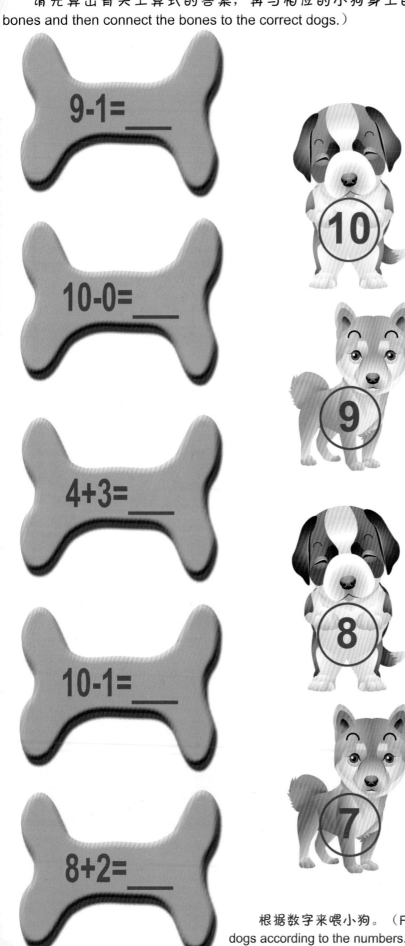
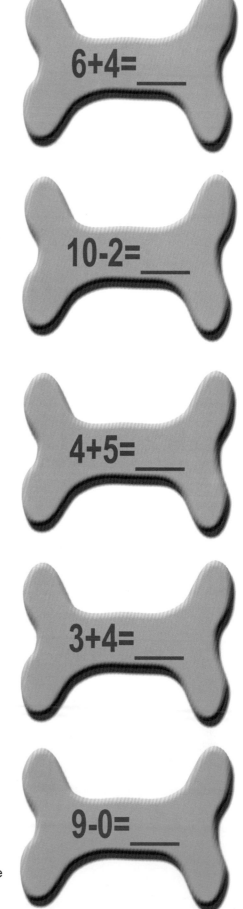

根据数字来喂小狗。（Feed the dogs according to the numbers.）

32

数学迷宫 — 奇数偶数路径

　　小老鼠要找食物吃，请帮这只小老鼠以"偶数 - 奇数 - 偶数 - 奇数"这样的模式从下图中的起点爬到终点吧，要小心不能遇到小猫哦。（Help this mouse creep from the start to the end by following a number path that goes even, odd, even, odd, etc. Don't meet the kittens.）

　　温馨提示：以 0、2、4、6、8 结尾的数字是偶数。以 1、3、5、7、9 结尾的数字是奇数。

数学游戏 — 九宫格

请在下面九宫格中的每一行、每一列和每个 3×3 方块中填入 1 到 9 的数字，使得九宫格中的每一行、每一列和每个 3×3 的小方块都包含 1 到 9 这几个数字，而且每个数字在每一行、每一列和每个 3×3 的小方块中仅仅出现一次。（Fill in the numbers 1 to 9 one time in each row, column, and 3x3 box within the following grids. Make sure that each row, column, and 3x3 box in the puzzle contains the numbers 1 to 9. If you do your job correctly, each of these numbers can appear only once in each row, column, and 3x3 box.）

温馨提示：请让爸爸妈妈帮你一起完成。

				3	6		9	5
3			9		4		7	
	9	7						
	3	1	6					
			2			1		
				9	5	4		
						2	5	
	1		4		7			8
4	6		5	9				

		1	8		6	9		
4		5						
	2		4	3				
9				7	2		5	
	8						1	
		2		9	8			6
				8	4		3	
						8		1
		4	3		2	7		

自制小书 — 早晨上幼儿园

在空白横线上填写时间，然后按照背面介绍的方法，制作一本关于早晨去幼儿园的迷你书，读一读。

（Fill in the times on the blank lines. Then make a mini-book about kindergarden according to the method on the next page. Read it.）

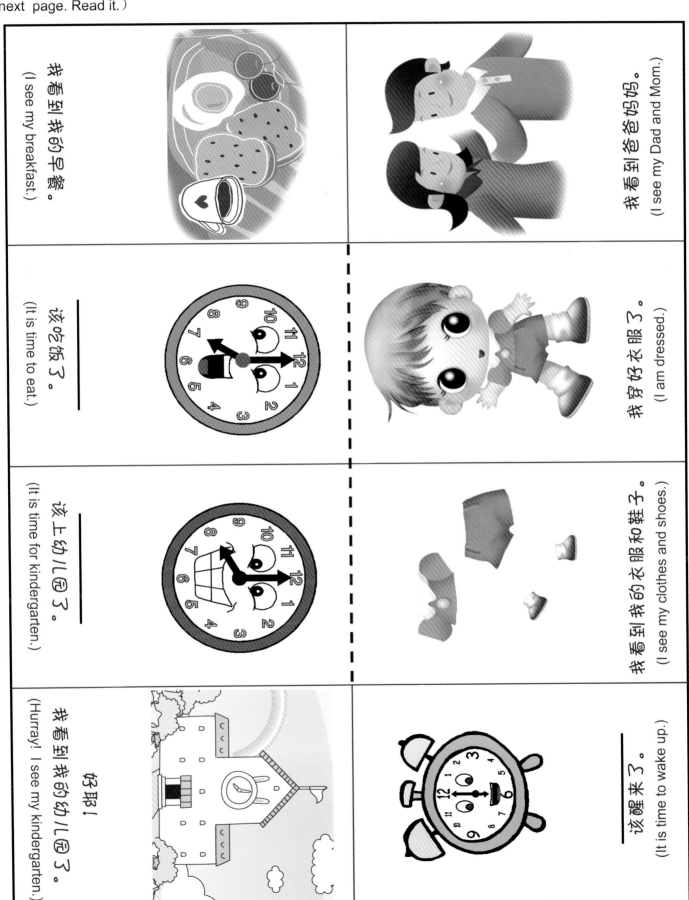

我看到我的早餐。
(I see my breakfast.)

我看到爸爸妈妈。
(I see my Dad and Mom.)

该吃饭了。
(It is time to eat.)

我穿好衣服了。
(I am dressed.)

该上幼儿园了。
(It is time for kindergarten.)

我看到我的衣服和鞋子。
(I see my clothes and shoes.)

好耶！
我看到我的幼儿园了。
(Hurray! I see my kindergarten.)

该醒来了。
(It is time to wake up.)

制作迷你书的方法
(How to Make Your Mini-Book)

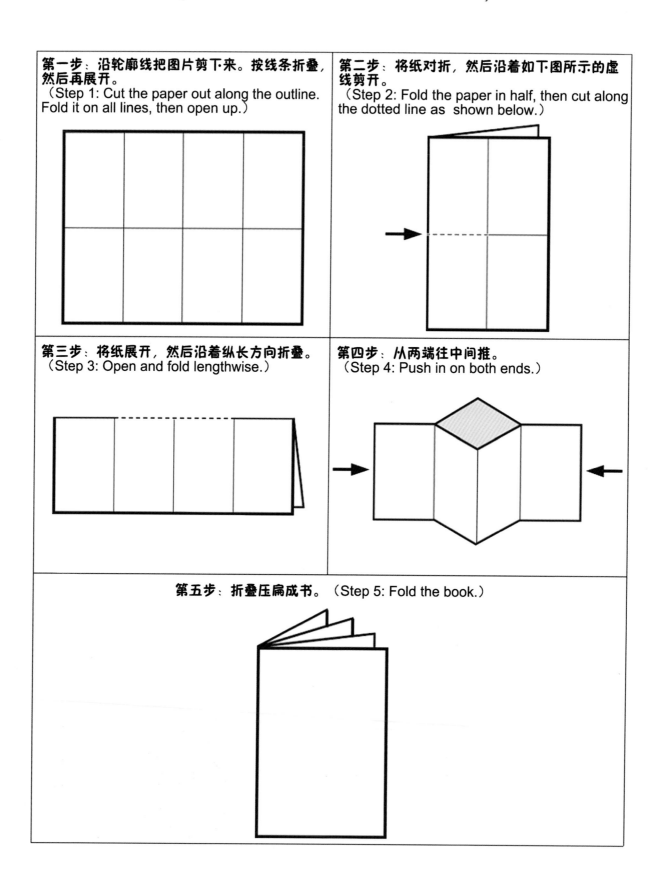

第一步：沿轮廓线把图片剪下来。按线条折叠，然后再展开。
（Step 1: Cut the paper out along the outline. Fold it on all lines, then open up.）

第二步：将纸对折，然后沿着如下图所示的虚线剪开。
（Step 2: Fold the paper in half, then cut along the dotted line as shown below.）

第三步：将纸展开，然后沿着纵长方向折叠。
（Step 3: Open and fold lengthwise.）

第四步：从两端往中间推。
（Step 4: Push in on both ends.）

第五步：折叠压扁成书。 （Step 5: Fold the book.）

仔细观看下面的图片一分钟，然后翻到下一页回答问题。（Look carefully at the picture for one minute, then turn the page to answer some questions.）

仔细观看下面的图片一分钟，然后翻到下一页回答问题。（Look carefully at the picture for one minute, then turn the page to answer some questions.）

前一页的图片内容你能记得多少？不许偷看哦！

（How many things can you remember from the picture on the previous page? No peeking!）

1. 有多少块积木？

（How many blocks are there?）

2. 有多少只蝴蝶？

（How many butterflies are there?）

3. 什么东西站在黑板顶上？

（What is on the top of the blackboard?）

4. 女孩裙子上是什么图案？

（What pattern is on the girl's dress?）

5. 男孩衣服上是什么图案？

（What pattern is on the boy's clothes?）

6. 积木上有一个什么字母？

（What letter is on one of the blocks?）

7. 积木上有一个什么数字？

（What number is on one of the blocks?）

8. 黑板上的算术题的答案是多少？

（What is the answer of the addition problem on the blackboard?）

9. 在图片右下角是什么？

（What is in the lower right corner of the picture?）

学会测量 — 物体的长度

剪下下面的测量尺。然后，用测量尺测量下面每个水平线条和交通工具的长度，将长度写在横线上。
（Cut out the measuring bar below. Then measure the length of the horizontal lines and vehicles below with your measuring bar and write the length on the lines.）

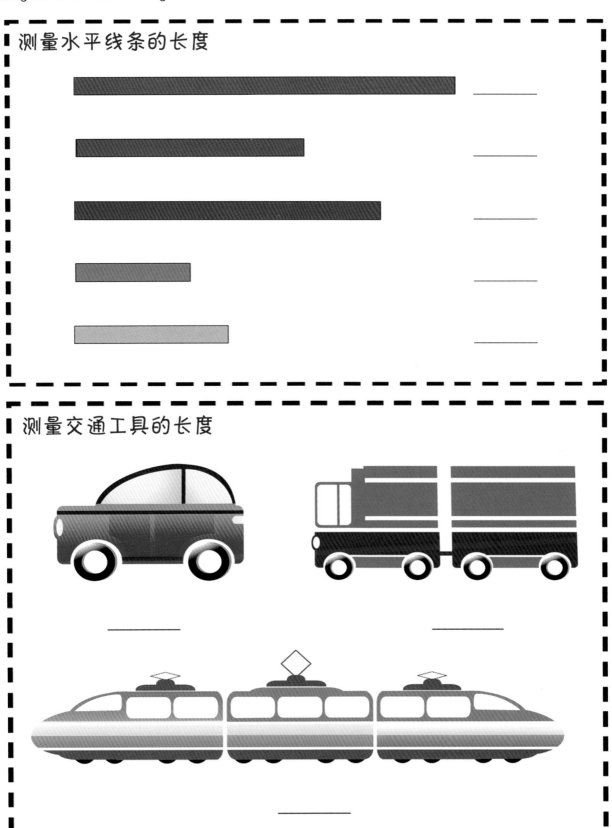

■ 测量水平线条的长度

■ 测量交通工具的长度

测量尺

1	2	3	4	5	6	7	8	9	10

请按照间隔 2 从 2 到 80 的顺序连接圆点，使之形成一只海豚。(Please connect the dots by twos from 2 to 80 to draw a dolphin.)

与众不同 — 闹钟上的数字

下面有一幅图与其他图不同，请将这幅图找出来。（Find the picture which is different from the others.）

数学迷宫 — 回家的路

算出下面数学加减法的答案。如果答案是1、2、3，那么就将格子涂成绿色。这样就可以画出小女孩回家的路。（Calculate the answers. If the answer is 1, 2, or 3, color the box green. Then make a path for the little girl back to her home.）

	9 −8	9 −6	5 −3	5 −2	3 +2
1 +3	6 −6	7 −2	4 +2	10 −8	2 +2
8 −6	5 −4	2 +0	6 −4	10 −9	4 +3
7 −4	3 +3	3 +5	10 −6	4 +5	9 −9
9 −6	5 +1	9 −7	7 −6	10 −7	1 +1
7 −5	7 −4	4 −3	6 +3	4 +6	10 −7
4 +4	5 +0	5 −5	2 +7	4 +3	

学会测量 — 物体的高度

剪下下面的测量尺。然后，用测量尺测量下面每棵树和建筑物的高度，将高度写在横线上。（Cut out the measuring bar below. Then measure the height of the trees and buildings below with your measuring bar and write the height on the lines.）

▌测量植物高度

▌测量建筑物高度

测量尺

1	2	3	4	5	6	7	8	9	10

43

数字有序 — 发现数字序列

在下面的数字格子中，数字序列 1、2、3、4、5、6 按顺序出现了一次。请找出来。此数字序列可能为对角线，甚至可能是倒序的。（The numbers 1, 2, 3, 4, 5, and 6 appear in order once in number grid below. The numbers could be diagonal. They might also appear backward!）

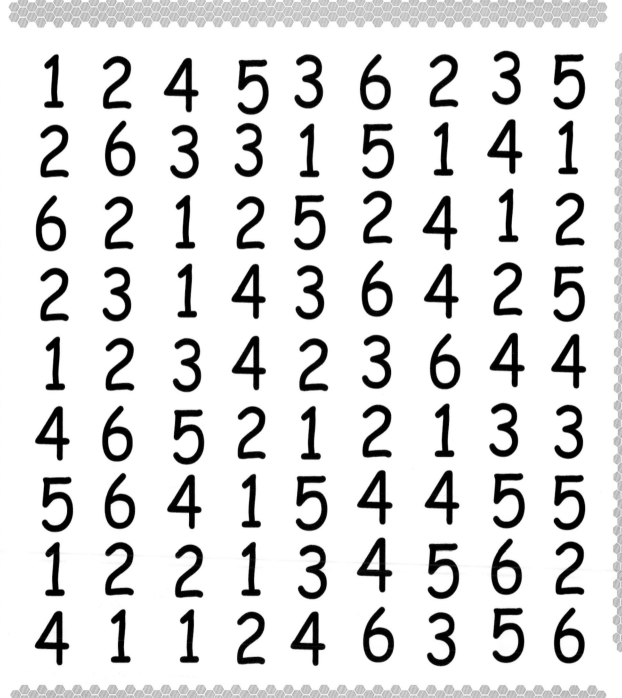

1	2	4	5	3	6	2	3	5
2	6	3	3	1	5	1	4	1
6	2	1	2	5	2	4	1	2
2	3	1	4	3	6	4	2	5
1	2	3	4	2	3	6	4	4
4	6	5	2	1	2	1	3	3
5	6	4	1	5	4	4	5	5
1	2	2	1	3	4	5	6	2
4	1	1	2	4	6	3	5	6

活动时间 — 上午的活动

请按照下面图片上的时钟指示，将正确的时间填写在空白处。（Complete the blanks with the right time indicated by the clocks in the pictures below.）

_____，我醒来。
(I wake up at _____.)

_____，我洗脸刷牙。
(I wash my face and brush my teeth at _____.)

_____，我穿衣服。
(I get dressed at _____.)

_____，我吃早餐。
(I have breakfast at _____.)

_____，我去上学。
(I go to school at _____.)

_____，开始上课。
(Classes start at _____.)

四分之一 —— 均分花朵

　　小熠想为他的四位朋友采花。请用两条直线将花分成相等的四束。每一束花应该有相同数量的三种花。（Xiao Yi wants to pick flowers for his four friends. Use two straight lines to divide the flowers below into four equal bunches. Each bunch should have the same amount of the three different kinds of flower.）

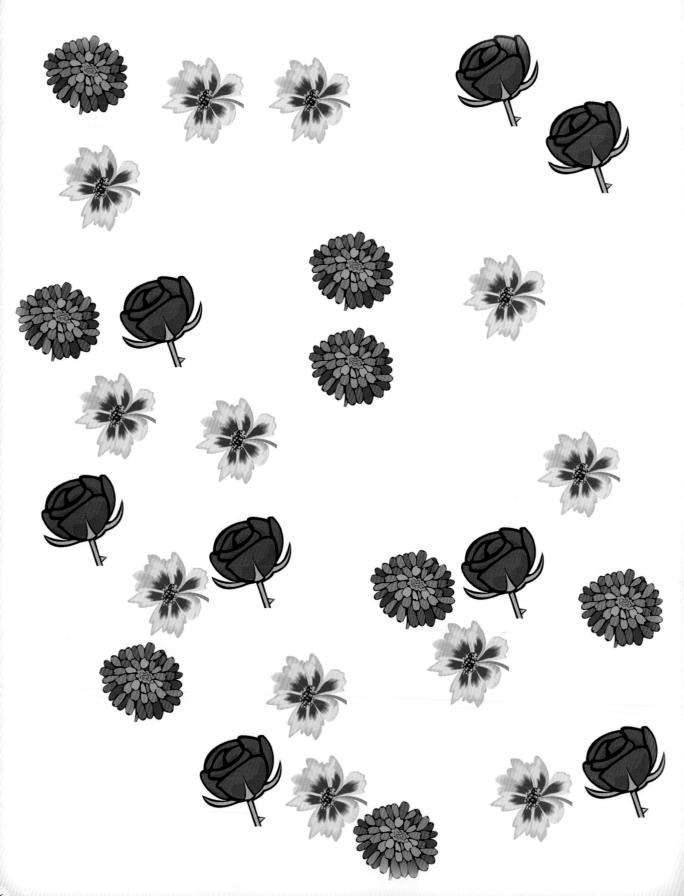

学买东西—买食品

从下图中勾选出尽可能少的纸币和硬币，获得需要购买每种食品的钱。（Tick the least number of bills and coins needed to buy each food.）

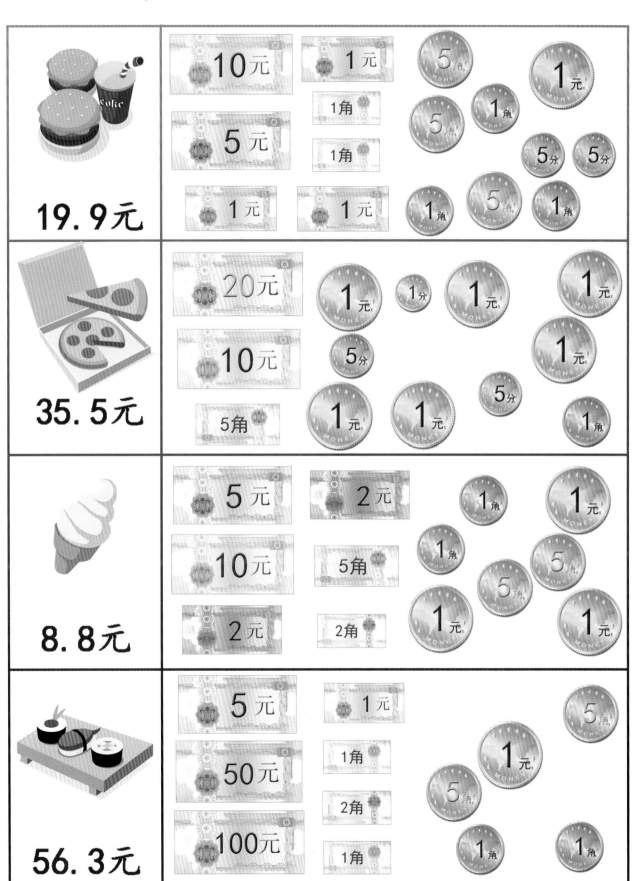

破译密码 — 学做加减法

阅读下面的问题，解答算术题。然后用数字与字母对应的关系表，将每个数字用字母代替，最后找出问题的答案。（Read the question, solve the arithmetic problems, and then use the number and letter table to substitute a letter for each number and find the answer to the question.）

温馨提示：请让爸爸妈妈帮你一起完成。

中国的首都是哪个城市？
(Which city is the capital of China?)

数字与字母对应关系

6=B 5=i
9=e 8=n
4=g 7=j

9 -3	7 +2	6 -1	10 -3	9 -4	5 +3	6 -2
答案						
字母 解密						

看图算术 — 花儿的加法

将答案填写在红框中。（Fill in the answer in the red box.）

3	+ 2	=
3	+ 3	=
2	+ 6	=

活动时间 — 中午、下午和晚上的活动

请按照下面图片上的时钟指示，将正确时间填写在空白处。（Complete the blanks with the right time indicated by the clock in the pictures below.）

_____，我吃午饭。
(I have lunch at _____.)

_____，我踢足球。
(I play soccer at _____.)

_____，我回家。
(I go home at _____.)

_____，我吃晚餐。
(I have dinner at _____.)

_____，我洗澡。
(I take a bath at _____.)

_____，我睡觉。
(I go to bed at _____.)

学不等式 —— 树叶的多与少

　　符号"＞"表示"大于"；符号"＜"表示"小于"。例如：3>2, 2<4。在下图横线中填入树叶的数量，然后圈选出正确的不等号。（Here, "＞" means "greater than", "＜" means "less than". For example: 3>2 and 2<4. Fill in the number of leaves, and then circle the correct inequality sign.）

　　温馨提示：请让爸爸妈妈帮你一起完成。

数字涂色 — 水上乐园

请按照数字对应的颜色给图片涂色。（Color the picture by number.）

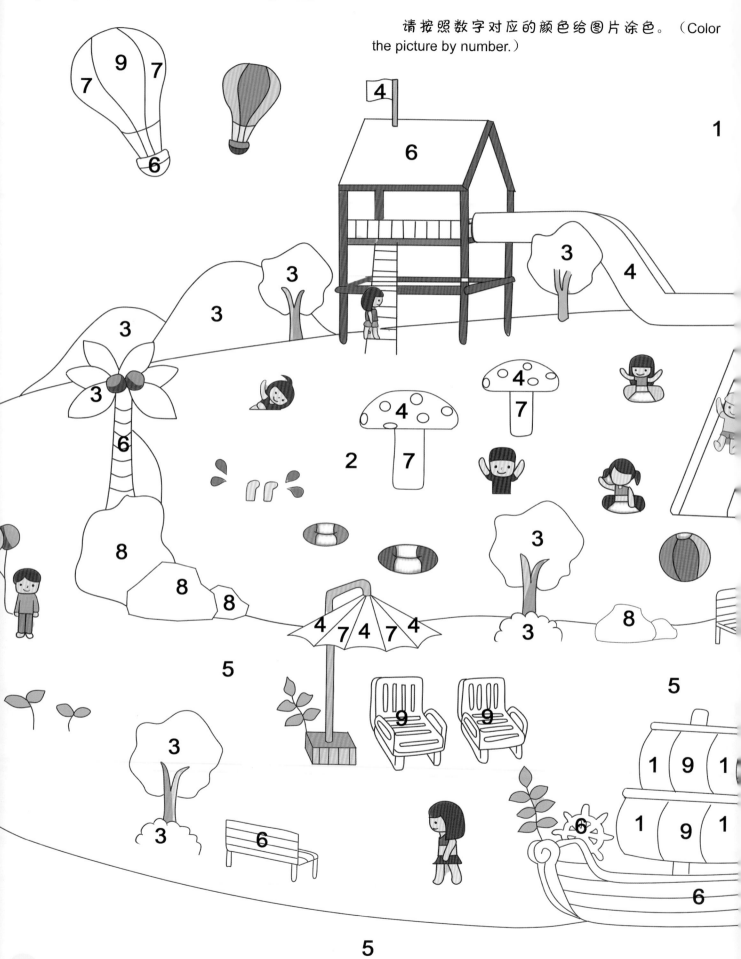

1= 浅蓝色（light blue） 2= 深蓝色（dark blue）
3= 绿色（green） 4= 红色（red）
5= 深黄色（dark yellow） 6= 棕褐色（brown）
7= 黄色（yellow） 8= 灰色（gray）
9= 紫色（purple）

数字有序 —— 制作三明治

你知道怎么制作三明治吗？看下面的图片，然后按照从 1 到 6 的顺序给句子填上数字。（Do you know how to make a sandwich? Look at the picture below, then number the sentences 1, 2, 3, 4, 5, and 6 in order.）

◯ 打开花生酱的罐子。
(Open the jar of peanut butter.)

◯ 吃光它。
(Eat it up.)

◯ 取出面包、花生酱和小刀。
(Take out the bread, peanut butter, and a knife.)

◯ 坐下来，大吃一口。 (Sit down and take a big bite.)

◯ 将面包切成两半。 (Cut the bread in two.)

◯ 将许多的花生酱涂在面包上。 (Put a lot of peanut butter on the bread.)

54

简单加减 — 小猫的食物

请先算出鱼身上算式的答案，再与相应的小猫身上的数字连线。（Calculate the answers on the fish and then connect the fish to the correct cats.）

10-4=___

3+3=___

5+3=___

8+0=___

9-3=___

2+5=___

6+3=___

10-3=___

9-2=___

4+2=___

9

8

7

6

根据数字来喂小猫。（Feed the cats according to the numbers.）

超市购物 — 买玩具

请按照每个玩具的价格在格子中填写数字，达到需要购买该玩具的金额。尽量少用硬币。（Fill in the grids with numbers to show the correct amount of money needed to buy each toy. Use as least number of coins as possible.）

温馨提示：请让爸爸妈妈帮你一起完成。

	10元	5元	1元	5角	1角	5分	1分
30元							
18元							
4元							
26元							
14.2元							
8.5元							
9.38元							
56.8元							
32.25元							
49.6元							
25.34元							

答案与提示

P1

P2

20	5	31	99	25	9
47	12	100	13	11	49
1	33	5	22	88	21
19	21	23	3	35	50
5	49	32	60	4	67
33	6	25	13	55	9
13	70	15	7	47	14
35	3	40	14	17	3
10	9	5	15	8	11
13	7	31	41	2	23

P3 5

P4

P5

1	2	3	4	5	6	7	8	9	10
11	12	13	14	15	16	17	18	19	20
21	22	23	24	25	26	27	28	29	30
31	32	33	34	35	36	37	38	39	40
41	42	43	44	45	46	47	48	49	50
51	52	53	54	55	56	57	58	59	60
61	62	63	64	65	66	67	68	69	70
71	72	73	74	75	76	77	78	79	80
81	82	83	84	85	86	87	88	89	90
91	92	93	94	95	96	97	98	99	100

P6 4、6、3、7、2、8、1、9、10

P7

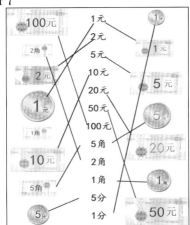

P8

P9

P10

P11

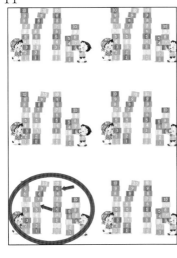

P12

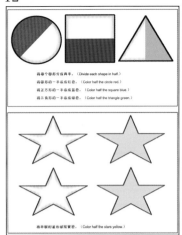

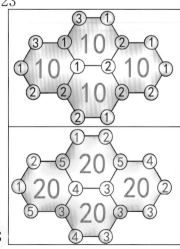

P13

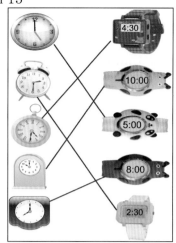

P14

P15

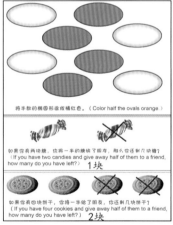

P16

P17

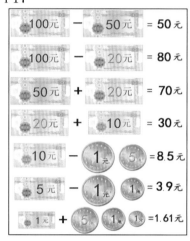

P18 1.6, 0.28, 1.15, 2.66

P19

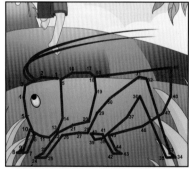

P20

I want to swim.
Come with me.
See you soon.

P21

P22

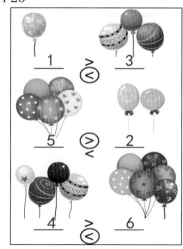

P23

P24

P25

58

P26

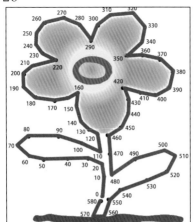

P27 0，2，1，3

P29

P28

3	0	0		3	3	2	1	2
+4	+2	+0		+1	+3	+3	+0	+4

答案	7	2	0		4	6	5	1	6
字母解密	t	h	e		p	a	n	d	a

P31 图片 B。红色圈选出的地方是不匹配之处。

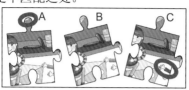

P32

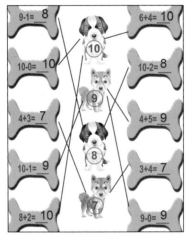

P33

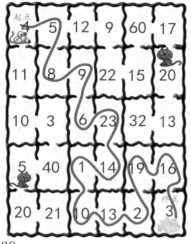

P34

1	2	4		3	6		8	9	5

P35 6:00，7:00，8:00

P38

1. 5 块积木（Five blocks）
2. 2 只蝴蝶（Two butterflies）
3. 一只鸟儿（A bird）
4. 圆点（Dots）
5. 条纹（Stripes）
6. A
7. 2（Two）
8. 8（Eight）
9. 一只狗（A dog）

P39

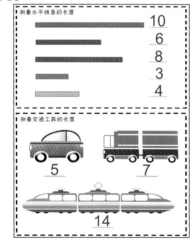

P40

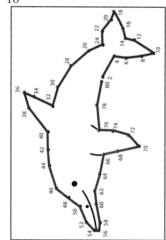

P41

P42

P43

P44

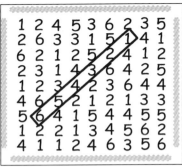

P45

P46

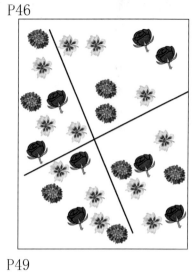

P47

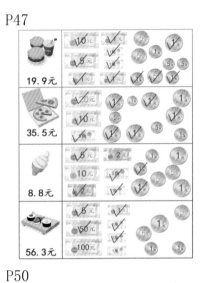

P48

P49

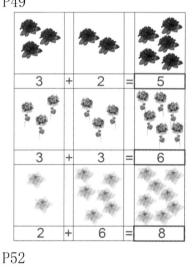

P50

P51

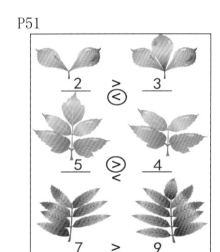

P52

P54

P55

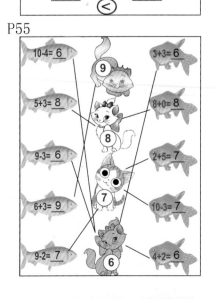

P56